EN_Electric-Motors

PLC Logics and HMI Screens

for

Electric Motors Automation

A pratical approach to electric motors monitoring and control
using IEC 61131 -3 Ladder Logic

AUTOMATION RECIPES - Volume 1

Rosario Cirrito

Copyright

All rights reserved. No part of this publication may be reproduced, stored in a retrieval system, or transmitted, in any form or by any means, electronic, mechanical, photocopying, recording or otherwise, without the prior permission of the author.

Every effort has been made to make this book as accurate as possible, however, there may be mistakes, both typo and in content. This content should be used as a guide, the result of a thirty-year experience of the author as a PLC - HMI - SCADA developer.

Suggestions, comments and requests for explanations or details are welcome, please send them to: author.rosario.cirrito@gmail.com

Note: This book contains many images. Since eReaders don't always display images well, I would like to provide you with the PDF file that contains this book and the images will be easier for you to see. To receive PDF version of this book, you just need to email me prove purchase of kindle version of the book from Amazon.com sending an email to author:rosario.cirrito@gmail.com. Upon receive of that, PDF version will be emailed to your mail box.

ISBN: 9781980690900
Independently published

EN_Electric-Motors

Synopsis

This booklet is the first of a series dedicated to automation recipes created with the PLC (Programmable Logic Controller) and HMI (Human Machine Interface) binomial. The series is aimed at an audience of readers with an elementary knowledge of PLC programming, eager to learn advanced solutions, extensively tested on real systems.

In modern computer programming, generally oriented to the development of "object-oriented" software, the developer strives, as much as possible, to resort to so-called "Design Patterns", standard solutions for frequently recurring problems. A design pattern describes a problem, particularly recurring in a given context, and then provide the heart of the solution to this problem. It is therefore possible to successfully reuse this solution, thousands and thousands of times, with the certainty of using an efficient and well-tested solution. Patterns can ultimately be considered as "elegantly formalized" best practices, which the programmer is quick to use to simultaneously achieve both an exponential decrease in development time and greater robustness and reliability of the generated code.

In the present series, which deals exclusively with development on PLC-HMI, the term "design pattern" has been replaced by the term "automation recipe" for an easier understanding by the non IT reader.

In the chapters of this book we will show in detail an automation recipe that can be reused in any PLC-HMI automation project that uses "electric motors". The recipe has also been optimized for operation with Scada supervision systems.

In detail the book illustrates the automation recipe for the automation of electric motors powered by three-phase alternating current.

The first section, dedicated to a brief introduction to the application domain, illustrates the physical structure of an electric motor and the main types of starting: direct at full voltage and with reduced voltage, obtained with a star - delta switching.

The second section deals with the development of combined software for both PLC and HMI.

The logic of the main functional block (UDFB), ElectricMotor, as well as the HMI monitoring and local control screens, are analyzed in detail. An auxiliary function block, the twin sequencer Mot2Seq, is then introduced, to explain the pratical usage of ElectricMotor in a real application for twin-pumps automation.

The HMI solutions have been extensively tested on the OCS, Operator Control System, manufactured by Horner Apg. OCS combines a Controller, Operator Interface, Network and I/O into a single

product. While the author, has been widely using Siemens, Allen Bradley, GE Fanuc PLCs he has focused the books of this series on the Horner OCSs because Horner provides Cscape, an integrated development environment, extremely easy to use and above all completely free.
All the logics, published in the book, have been developed using the IEC61131-3 compliant Ladder language; therefore it is extremely easy to migrate them on almost all the PLCs of other manufacturers.
The same applies to HMI screens whose graphic controls are very similar on the different equipment offered on the market.
The reader who already has experience with other manufacturers' equipment can therefore continue to use what he knows best.

Contents

Cover
Title
Copyright
Synopsis

1. The application domain
1.1 The electric motor

2. The application PLC - HMI software
2.1 The modular programming and the memory mapping
2.2 The Electric Motor function block
2.3 The Electric Motor Human Machine Interface
2.4 The twin sequencer function block
2.5 The twin sequencer Human Machine Interface

3. Summary

1.1 The Electric Motor

Most of the plant components such as pumps, fans and compressors are coupled to three-phase asynchronous electric motors with wound rotor or squirrel cage. It is a very simple and robust rotating machine set up in the last century, independently, by Galileo Ferraris and Nikola Tesla.

The three-phase asynchronous motor consists of two distinct parts: the stator (fixed) and the rotor (rotating). The operating principle is based on the production of a rotating magnetic field obtained by feeding the fixed coils of the stator with three-phase alternating current.

The stator consists of a metal casing on which a crown of special steel sheets is fixed, provided with special grooves. The electrical copper windings are distributed within these grooves. Their assembly is called "stator winding". The rotor, on the other hand, constitutes the moving part of the engine; it is placed inside the stator and consists of a cylinder, made of stacked steel sheets, fitted on a cylindrical shaft. The connection of the stator windings of a three-phase asynchronous motor can be carried out in two ways: star or delta.

The direct start-up feeds the motor with voltages equal to those connected between phase and phase. This ensures a high starting torque, but on the other hand, the motor absorbs a considerable current peak, about 4-8 times, at the start-up, the nominal operating current. The engine develops all its torque for the benefit of a shorter acceleration time, however high inrush currents create voltage drops on power lines that may not be well tolerated by other loads connected to the network.

The star-delta switching, on the other hand, allows the motor to start up at reduced voltage, corresponding to the line voltage between phase and neutral line. It is obtained by initially connecting the star motor windings, and then switching, after about ten seconds, to full-voltage work with the delta connection. The main advantage of this starting system is a lower value of the current absorbed at start-up. The star-delta switching is preferably obtained directly with an electromechanical timer, and not by means of the PLC logic. The reason is that by doing so, even in the event of a PLC fault, it is still possible to start the star-delta, by operating manually on the AUT-0-MAN selector located on the front-panel.

The PLC must be programmed, in any case, to control the time-delayed starting of the individual motors in order to limit the global current absorption from the power line and the consequent voltage drop.

The single-circuit diagrams of an electrical control panel provide a power section and an auxiliary circuits section.
The power section of a control panel for three electric pumps is shown in figure 1.1.1:

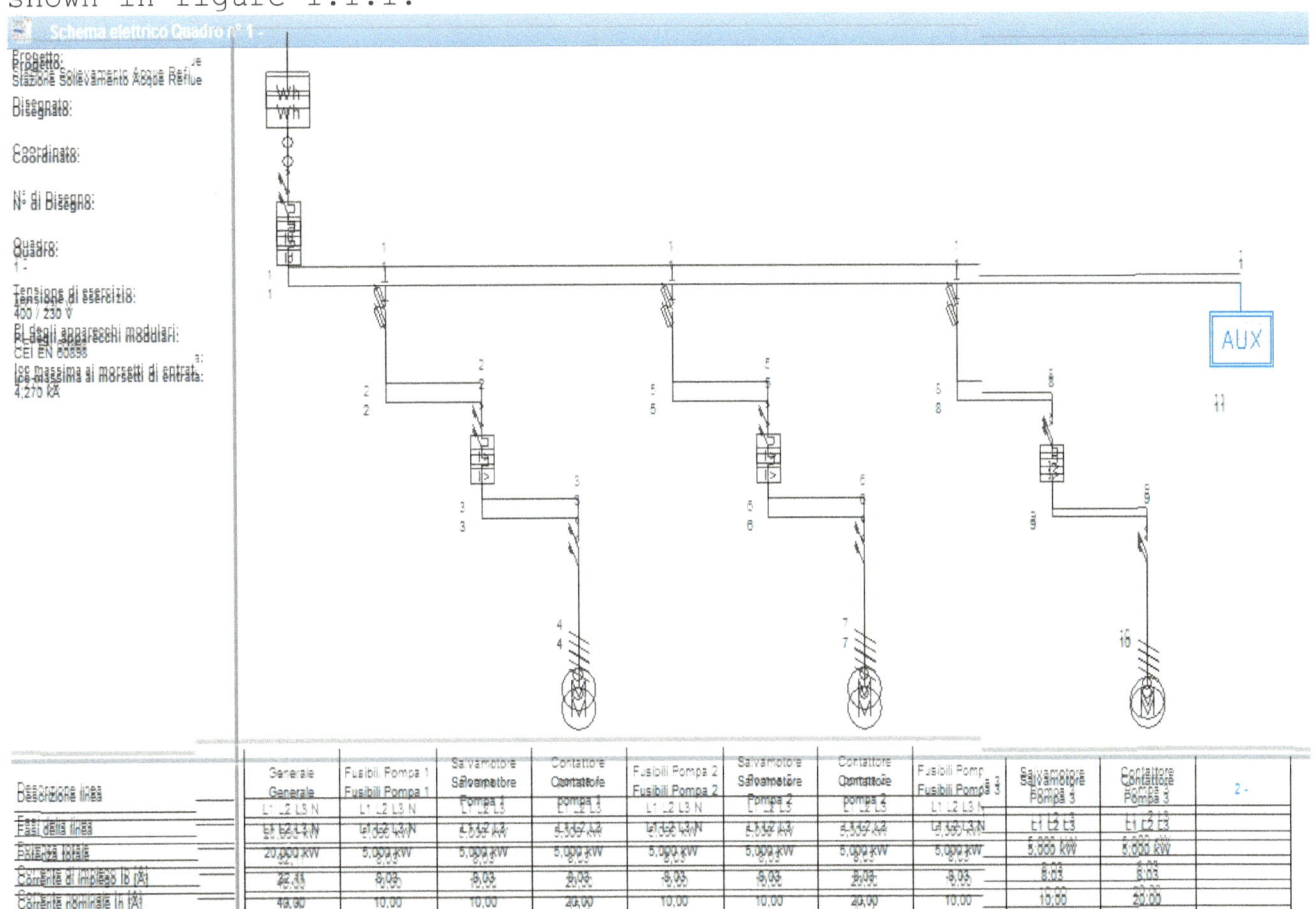

The auxiliary circuits section must show the correct interfacing between the electromechanical contactors of the power section and the PLC.
Figure 2.2.2 shows the auxiliary section related to the first pump. The following ones for the remaining pumps can easily be duplicated from the first.

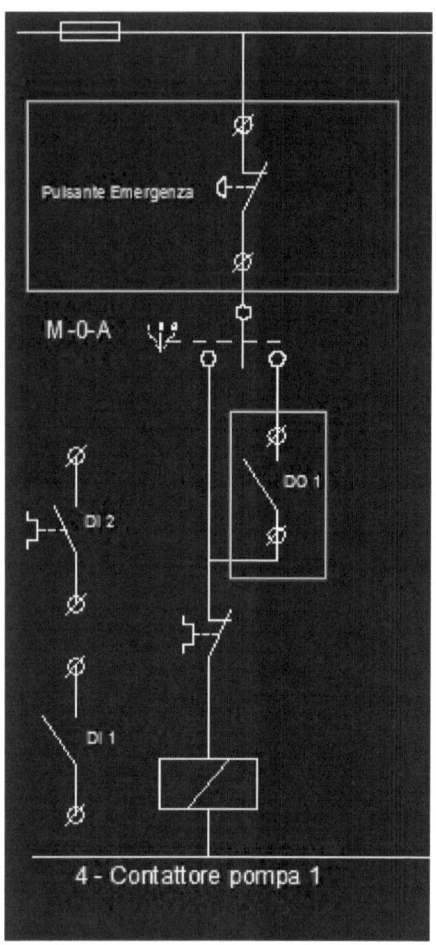

Proceeding from top to bottom we have the emergency button wired, for intrinsic safety, as a normally closed NC contact. The rectangular frame shows us that the emergency button is located outside the power panel. The crossed circles indicate that its wiring must be returned to the appropriate interface terminal to the automation panel. If the emergency button is not present on the system it will be enough to jumper the relative terminals.
A MAN-0-AUT three-position selector is wired in series with the emergency button. The automatic branch is intercepted by the contact, normally open NA, associated as we will see with a digital output, DO, which stands for Digital Output, of the PLC. Also this contact is inside a rectangular box to remind us that it is placed outside the control panel; in fact it is located within the automation framework. This contact must also be reported in the interface terminal board. The NC protection of the thermal relay is wired in series to the branches of the MAN-0-AUT selector and, subsequently, the coil of the pump contactor. The pump can therefore be started only with the simultaneous consent of the emergency button, of one of the two branches of the selector and of the thermal protection.

There are then two contacts on the left side of the diagram, whose wiring should be reported exclusively on the interface terminal board. The lower one is an NO contact obtained from the pump contactor; we will need to have the pump ON feedback on a first digital input of the PLC. The one placed above is also an NO contact obtained from the thermal relay; we will need to inform the PLC, thanks to a second digital input distinct from the first, that the manual reset thermal protection has intervened.

2.1 The modular programming and the memory mapping

The types of languages provided by the IEC 61131-3 standard can be considered as low-medium level programming languages.
This standard was developed to ensure a certain portability of the programs between PLCs from different suppliers.
Its greatest merit consists in being oriented to the modular development of control applications thus allowing the overall logic to be divided into subroutines recalled cyclically by a single main program.
The individual subroutines can in turn recall standard functional blocks provided by the language or even functional blocks UDFB, acronym of User Defined Function Block, expressly developed by the user.
A function block groups an algorithm and a set of private data so it is parameterized with reference to the input and output variables; this allows the developer to create multiple instances of each block belonging to the same automation system.
We will develop this important concept in greater detail in the following paragraphs.
To program subprograms and function blocks, the standard allows the developer to use one of the following five languages:
1) Ladder Diagram (LD)
2) Instruction List (IL)
3) Function Block Diagram (FBD)
4) Structured Text (ST)
5) Sequential Function Chart (SFC)

The choice is dictated by personal preferences or by the specific professional background of the programmer.
The modular decomposition of the application is clearly visible in Figure 2.1.1 which shows the structure of the various components typical of a project realized with the Horner CScape development environment:

EN_Electric-Motors

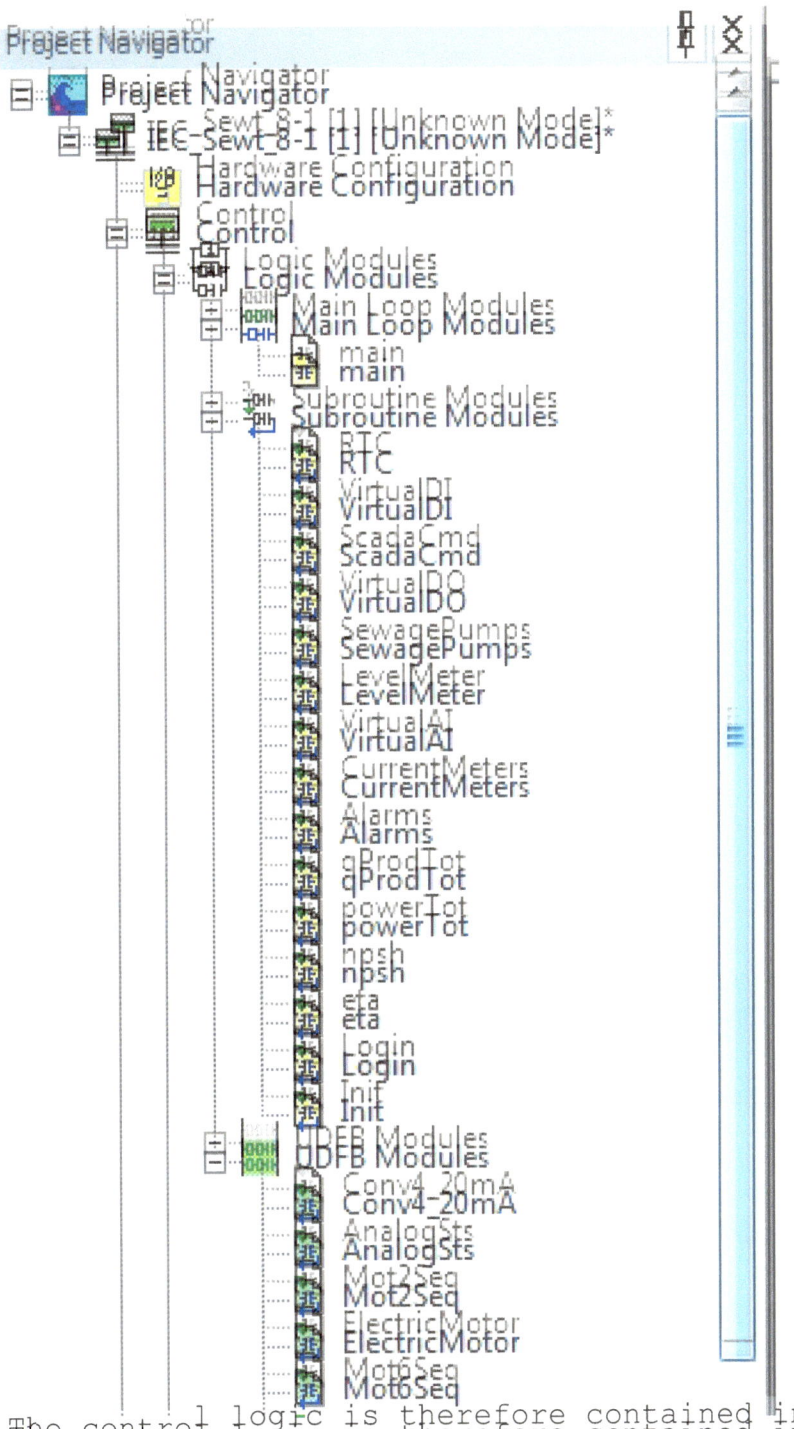

The control logic is therefore contained in a series of "Logic Modules": in summary, at the top of the hierarchy of modules we have the Main Loop Modules that contain at least one main module, which runs cyclically. The main program calls in sequence, one at a time, the various "Subroutine Modules" which in turn can recall the functional blocks several times, but with different parameters each

time. The latter can be both the standard function blocks, already provided by the language for the general purpose logical instructions, or the specific user-defined blocks, the "UDFB Modules".
The PLC control logic as well as the HMI display logic is however developed on a PC, almost exclusively in the Windows environment. The relevant source file is saved, periodically and after complete modifications, on the PC hard disk.
Whatever the hardware being used, PC or PLC, a set of RAM working memories is always needed both to store the program instructions and to save, at each scan cycle, the data of the dynamic variables.
Today's PC generally has a RAM of 4-8 GBytes, while the PLC requires far more modest memories, from 256 kB to 1MB, to store control logic of even particularly complex systems, as well as a few thousand internal variables.
High-level PC languages use Short, Byte, Integer, Long, Float, Double primitive variables that occupy from 8 to 64 bits of memory; the data type most frequently used by the PLC is instead the Word, composed of 16 bits, also called register (% R). A single Word or Register, having 16 bits in total, thanks to the binary numbering system, can represent Integers with a sign between -32768 and + 32767 or without sign in the range 0 and 65365.
When it is necessary to represent integers of higher value or real number, a 32-bit representation is used, obtained by merging two adjacent 16-bit registers.
A 16-bit register can also be used to aggregate the binary state of logical bits, each of which occupies one bit, in groups of 16. This packed solution is particularly compact and efficient especially when these variables are transmitted to Scada systems or transferred on the network from one PLC to another.
The individual bits of the Boolean variables then become individually accessible at address% Rx.y with x, index of the register, and y index of the bit, between 1 and 16: so %R1.5 will indicate the bit 5 of the register 1. Boolean binary values can in any case be stored also as retentive variables of type% M or non-retentive of type% T.
In addition to storing integers, real numbers and packed bits, the% R registers are also used to store enumerations of machine and sensor states that can be associated with predefined text strings in the HMI interface screens. We will show such use when we take care of displaying dynamic texts in the HMI panel.
A memory representation of a physical quantity, acquired in real time, is, for example, a pressure that at a certain moment takes on a value equal to 8.95 bar. In this case we can represent it as a 32 bit real value, using two consecutive registers, for example %R201 and

%R202; or as an integer value, equal to 895, with occupation of a single 16-bit register, for example% R200.

This second mode allows us to store the actual data in half the memory space, which is especially important when the data must be sent to a supervision system or another controller along a serial line that is not too fast; but this approach has the drawback that the correct representation of the displayed format must always be managed, by PLC logic and HMI configuration, keeping in mind how many decimal digits are to be taken into account.

An example of a register containing an enumeration of dynamic texts is constituted by the status register of a pump whose value can vary, in real time, within a certain set of logic precoded states stored in tabular form within the HMI device, as shown in figure 2.1.2.

Value	Text
0	???
1	ON
3	ON_SEL
4	OFF
5	REM_0
6	LOC_0
18	ALARM
32	INIBIT
64	INTERD
130	FDBACK

The integers shown in the Value column correspond to the logical states shown in the Text column. The latter can therefore be displayed in a text field inside the graphic pages of the operator panel associated with the PLC.

2.2 The Electric Motor function block

Problem

The ElectricMotor function block satisfies the primary requirement to control the start-up of an electric motor, either automatically or by a local operator panel or by a remote supervision system.
This block is used for monitoring and control of machinery driven by electric motors such as for example: electric pumps, electric fans, compressors, etc. which are the main components of technological systems, both industrial and residential.

Solution

The **monitoring** logic provides for the real-time acquisition of the operating status of the machinery. By appropriately interrogating some input bits and suitable internal variables, it is checked whether the device is running, if it is in normal stop waiting to start again or if it is stopped due to the presence of an alarm condition or inhibition on starting. Monitoring is integrated by the automatic calculation of operating hours and number of starts; two parameters very important for correct preventive maintenance. Then the "feedback not ok" failure status check is carried out which occurs when the machinery, although commanded to start, does not start running properly within a predetermined time, and lastly the manual run check which occurs when the engine is started by the manual branch of the electromechanical selector MAN-O-AUT present on the power front panel. The actuation of this selector by-pass obviously any automatic control of the PLC, so it is extremely important that the controller can detect this operating condition in order to inform of the event the internal control strategies.

The **control** logic consists essentially in activating the electric motor contactor coil, once a machine start request is received from a higher level management strategy. The start will be commanded provided that the intrinsic safety of the machinery is respected, such as the consent of the thermal protection relay, of the eventual emergency button and, when present, of any external security switch.
A pre-programmed restart inhibition time is also respected at each machine stop. This block is dictated by the need to avoid too frequent starts that could lead to overheating of the electric motor due to the high currents absorbed during start-up.
The management strategies of control, better detailed in the following chapters, can be twin starting strategies, which alternatively command one unit while the other remains in reserve, or

multiple start / stop strategies that operate, according to the needs of the plant; one or more components of a set of pumps or compressors installed in parallel.

Map of local variables

The map of the variables, codified according to the IEC61131-3 standard, used by the functional block is shown in the table shown in figure 2.2.1.

The input variables are:

ElectricMotor		
RemCmd	INT	IN
LocCmd	INT	IN
ExtLock	BOOL	IN
On	BOOL	IN
ThermProt	BOOL	IN
Go	BOOL	IN
SetInterdTim...	INT	IN
iNr	INT	IN
iHh	INT	IN
rhReset	BOOL	IN

The input variable **RemCmd** is a 16-bit register of the PLC that stores the value of the remote command coming from the Scada system. As we will see later, a convention has been adopted whereby the value 0 of this register causes operation to follow the automatic logic of the PLC control strategy. The value 1 of this register means that the operator of the Scada system wants to force the remote shutdown of the machinery. The value 2 of this register means that the operator of the Scada system wants to force the remote start of the machine.

The input variable **LocCmd** is another register of the PLC that stores the value of the local command coming from the HMI keypad. As we will see further here, it has been decided that the value 0 of this register causes the operation to follow the logic of the RemCmd command illustrated above; the value 1 of this register means that the local operator wants to force the machine to stop locally regardless of the value of RemCmd. It can thus make exclusions of machinery that are known to be damaged pending subsequent repair. In the case of maintenance checks to avoid unwanted starting of the machine, the operator will preferably act directly on the AUT=0=MAN selector of the power panel or in the most dangerous cases and for greater safety he will open the power fuses. The value 2 of the LocCmd register means that the HMI operator wants to force the machine to start locally.

The subsequent input variables On and ThermProt are boolean variables related to the previous digital inputs described in the previous chapter 1.1.

The ExtLock input variable is used to acquire the status of any external security switch.
The input variable Go comes from the sequencer control strategy that will enable or not the starting according to the system requests. In the case of water pressurization sets, it will be the control strategy of the discharge pressure, in the case of wastewater lifting stations it will be that of the level inside the tank.
The SetInterdTimer input variable is a PLC register in which the setting value of the inhibit timer is stored. This prevents the engine from being restarted for a certain number of seconds after it is switched off in order to allow it to be cooled down. The value to be set can easily be calculated by knowing the maximum number of starts / hours indicated by the electric motor manufacturer. For example, in the case of 10 starts / hour, the set in question will be set to 3600/10 = 360 s.
Finally, the iNr and iHH variables contain the registers of the number of starts and the number of operating hours of the machine. The number of starts is automatically reset when the value of 30,000 is exceeded (a 16-bit register can only store integers up to 32,767) while the hours are reset when the value of 30,000 is exceeded or when the last rHReset boolean variable is activated.
Let's move on to the output variables of the functional block which, as shown in figure 2.2.2, are:

Name	Type	Direction
Status	INT	OUT
oNr	INT	OUT
oHh	INT	OUT
FbNOk	BOOL	OUT
Start	BOOL	OUT
Ready	BOOL	OUT

Let's start with the Boolean variables: Start is correlated with the digital output that drives the machine. The other two Boolean FbNOk and Ready variables are mainly used as internal indications for control strategies. In the case we have two pumps, if the first one is not ready (Ready) while the second one is, the control strategy would give the Go to the second rather than to the first.
FbNOk (feedback not ok) indicates that the On signal did not become true even though the Start was commanded. This indicates any interruption in the chain of command for which a maintenance job is required.
The variables oNr and oHH contain the updated values of iNr and iHH after the execution of the function block.

The Status variable contains the PLC register that stores the possible logical states of the machine. The values that this register can assume will be illustrated in detail below.
Finally, the variables inside the functional block, as shown in figure 2.2.3, are:

Variable	Type
CtTFB	INT
TFB	TON1s
RemAut	BOOL
LocScd	BOOL
RemStart	BOOL
RemZero	BOOL
LocStart	BOOL
LocZero	BOOL
oldOn	BOOL
Interd	BOOL
CtInterd	INT
InterdTmr	TON1s
secCounter	INT
ionsec	BOOL
ionhours	BOOL

Their function will be described in detail in the next paragraph. The control logic is contained in the UDFB module called Electric Motor. An example of EM_EP1 instance, for controlling the EP1 pump, is shown in figure 2.2.4:

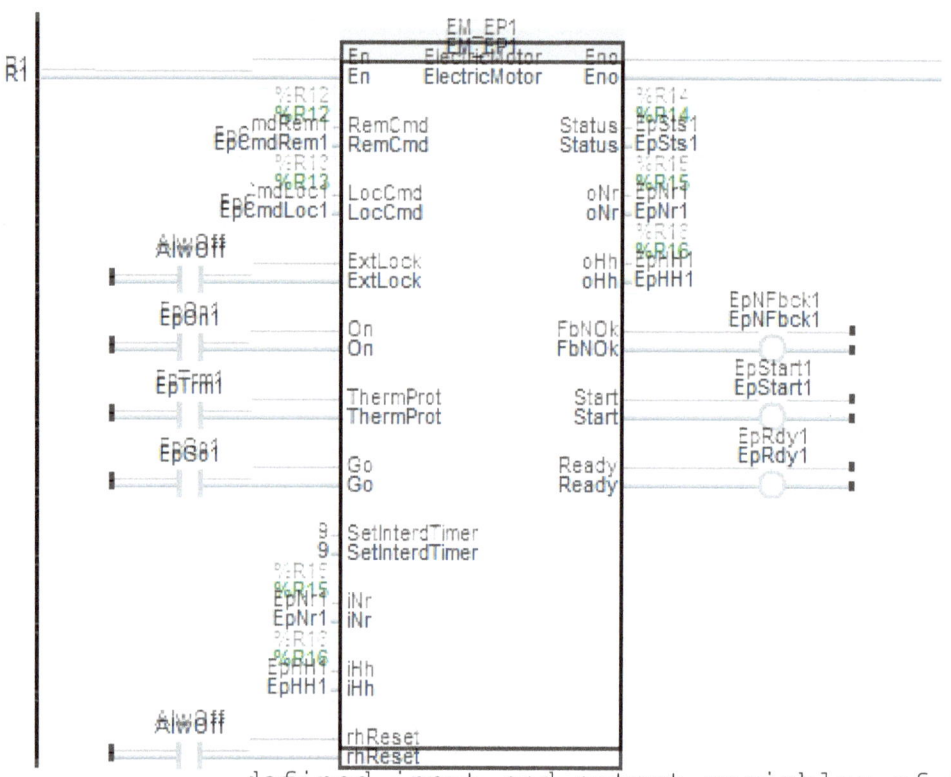

The previously defined input and output variables of the module are linked to the process variables of the specific controlled machinery: The EpCmdRem1 register declared as a **retentive** variable for the electric pump 1 is associated to the RemCmd, EpCmdLoc1 to LocCmd and so on.

The ExtLoc parameter is not used in the this example so it is then associated with the system variable AlwOff (Always Off).

The internal instructions of the block begin with the management of the interdiction timer contained in the R1-R4 rungs, as shown in figure 2.2.5:

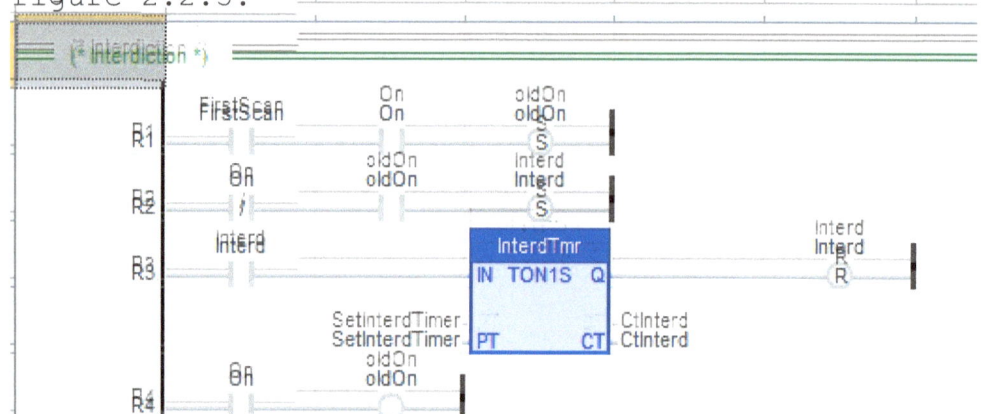

As we can see, rung R1 contains the series between the FirstScan and On contacts. The FirstScan system boolean variable is only true when

the PLC is first started. If the Boolean variable linked to the On input is also true, the oldOn internal variable is also set to true.
In the next rung R2 the series between the contacts of the boolean variable denied On and of the oldOn variable sets the variable Interd.
In the third rung R3 this variable activates the interdiction timer when the time set in the SetInterdTimer register elapses. Remember that the timer function is to avoid overheating of the motor in the event of too frequent restarts.
Rung R4 forces the update of the boolean variable oldOn based on the actual value of On.
The next instruction rung R5 sets the output variable FbNok (Feedback not ok) corresponding to the status message with a delay of 10 seconds after the occurrence of the logical series between the Start command and the lack of the On signal as shown in figure 2.2.6:

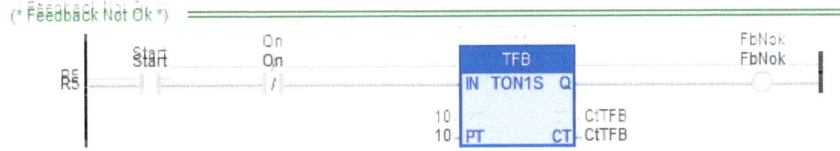

The output variable FbNok (feedback not ok) will be used by a sequence strategy, of a higher level, to start, alternatively, the other machines that may be available.
The module can be easily modified by the user if he wants to make programmable the parameter of 10 seconds otherwise set as a fixed value.
The subsequent rungs R6-R11 set internal Boolean variables according to the values of the RemCmd and LocCmd registers as shown in figure 2.2.7 - 2.2.9. If the registers are equal to 0, the RemAut variables (remote operation set to automatic) and LocScd (local operation set to Scada) are activated.

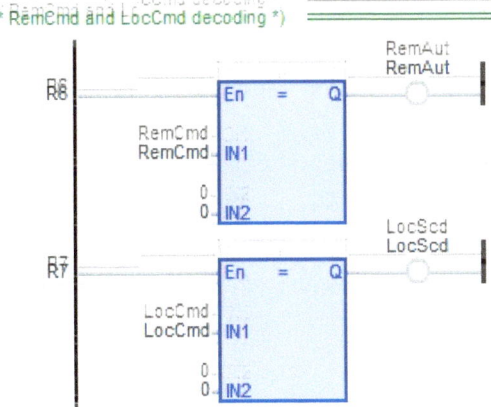

If the registers are equal to 1, the RemZero variables (remote operation set to zero) and LocScd (local operation set to zero) are activated respectively.

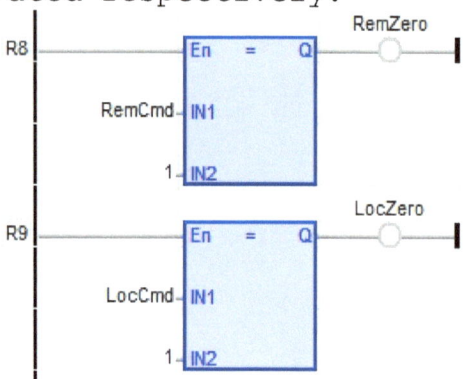

If the registers are greater than 1, the RemStart variables (remote start) and LocStart (local start) are activated.

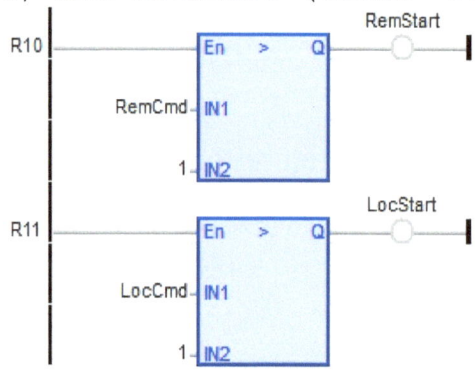

The next rung R12 is used to set the output variable Ready (ready to start) as shown in figure 2.2.10. This variable is also used by the machinery sequence strategy.

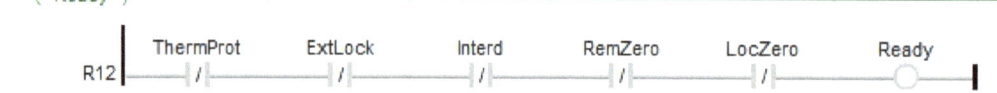

The Ready variable is set to true in the absence of thermal protection, external block, interdiction and both local and remote zeros. Rung R13 is used to set the output variable Start used to command the start of the machine as shown in figure 2.2.11.

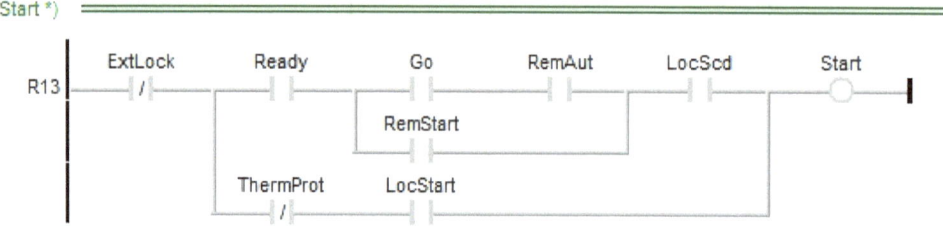

Following the logical flow of the direct branch we observe that the Start variable is activated in the absence of external block and in the presence of the Ready variable (ready to start,) of the Go variable set by the machinery sequence strategy, and of the remote operation RemAut variables in automatic and LocScd local operation allowed by the Scada.
The parallel branch with the RemStart contact is used to start the machine remotely, from the Scada system, by-passing the series of Go and RemAut contacts.
The lower branch with the LocStart contact is used to immediately start the local machinery, provided that the ExtLoc and ThermProt safety devices have not intervened.
The following rungs R14 and R15 manage the count of the number of starts and its reset when the value of 30,000 is exceeded as shown in figure 2.2.12. Each time the engine starts, marked by the positive transition, from 0 to 1, of the boolean variable On, the start counter is increased by 1. It will be reset to the initial value of 1 when the threshold value of 30,000 is exceeded.

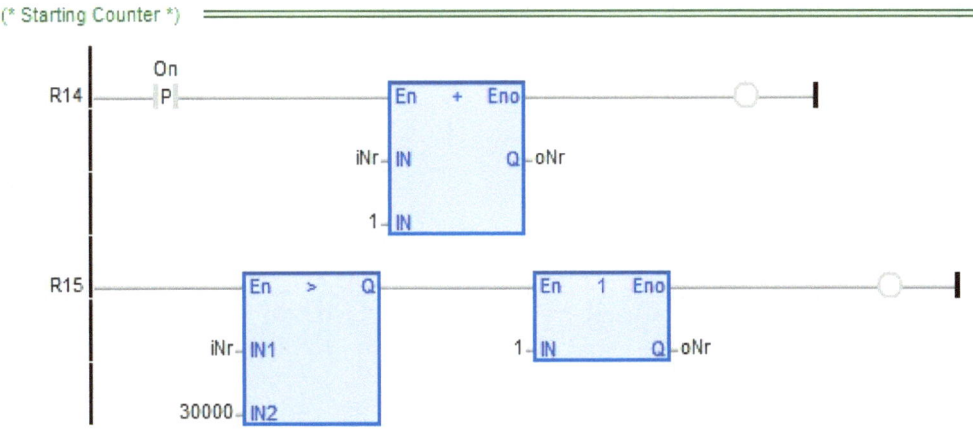

The R16-R22 rungs manage the count of the operating hours and the reset when the value of 30.000 is exceeded or when the rhReset input parameter is activated as shown in figure 2.2.12. The seconds spent in the On condition are totalized in the secCounter variable and when the value of 3600 is exceeded, the ionhours impulsive flag is set while the secCounter counter is reset as shown in figure 2.2.13.

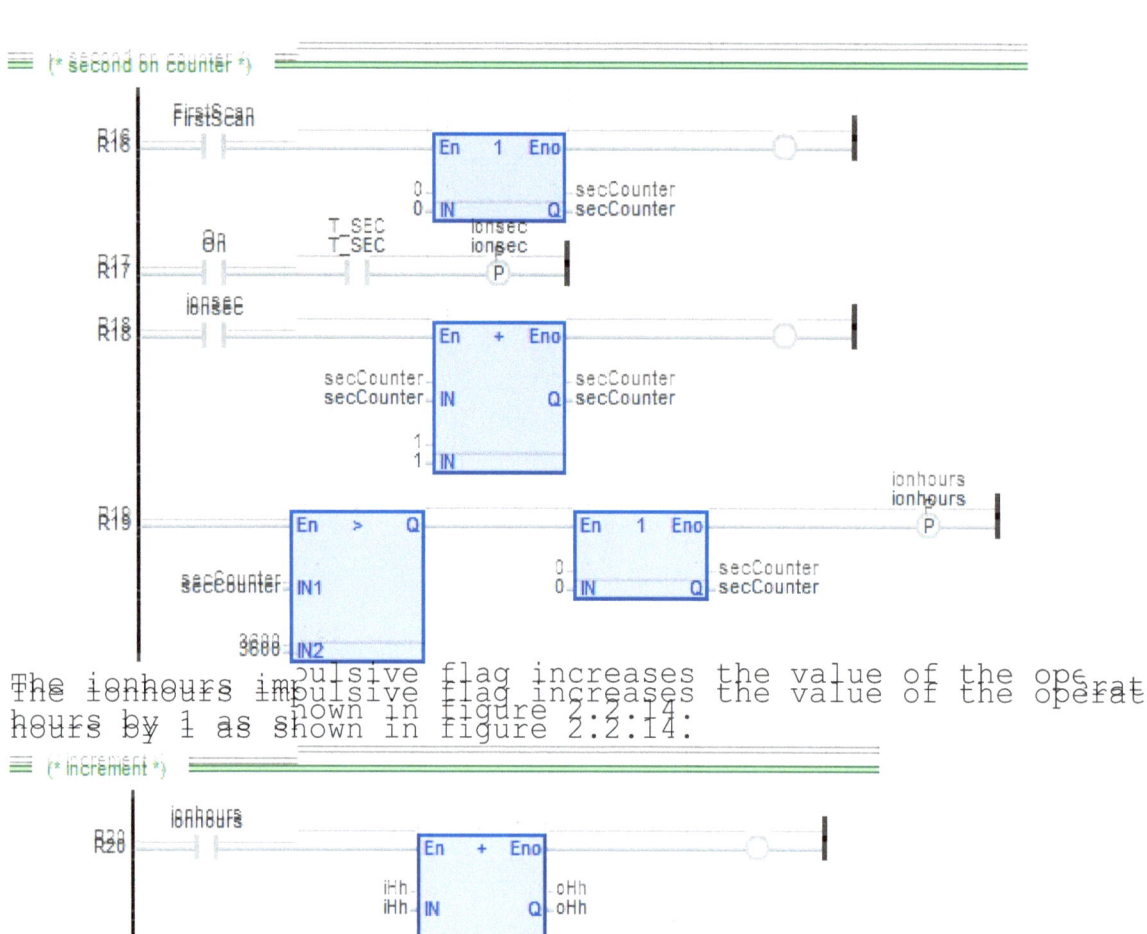

The ionhours impulsive flag increases the value of the operating hours by 1 as shown in figure 2.2.14:

When the value of 30,000 is exceeded, the hours counter is reset to 1 while the seconds counter is reset to 0 as shown in figure 2.2.15. If the rhReset variable is true, both the number of starts and the operating hours are reset. This possibility can be used in case of replacement of a damaged engine with a new one.

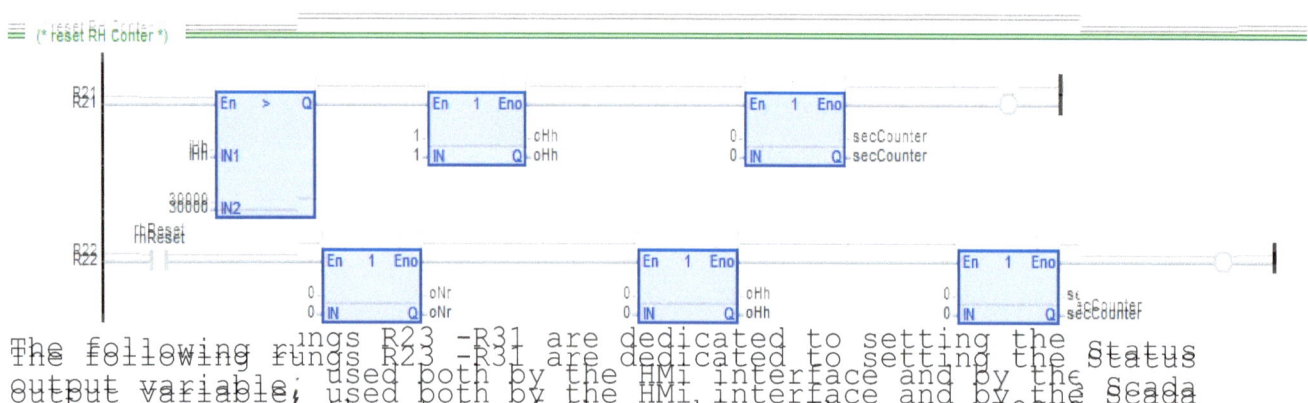

The following rungs R23 -R31 are dedicated to setting the Status output variable, used both by the HMi interface and by the Scada system to decode the state of the machine. The rungs R23-24, as shown in figure 2.2.16, manage the states of On (value 1) and On by selector on the front of the frame (value of 3). This last state is in fact related to the presence of the On signal even in the absence of control by the PLC:

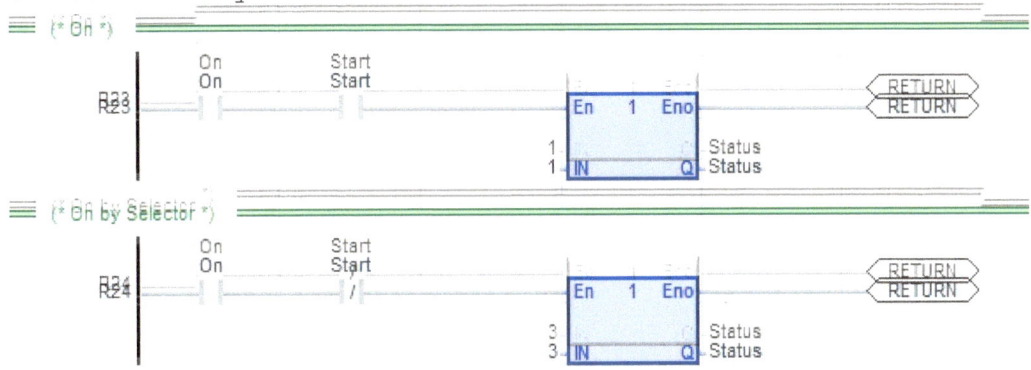

Rungs R25=31 manage the states of Off. If the machine is Off, the value 4 is set in normal conditions; if instead it is set to zero remotely, the value 5 is set and if the value 6 is zero in local, as shown in figure 2.2.17:

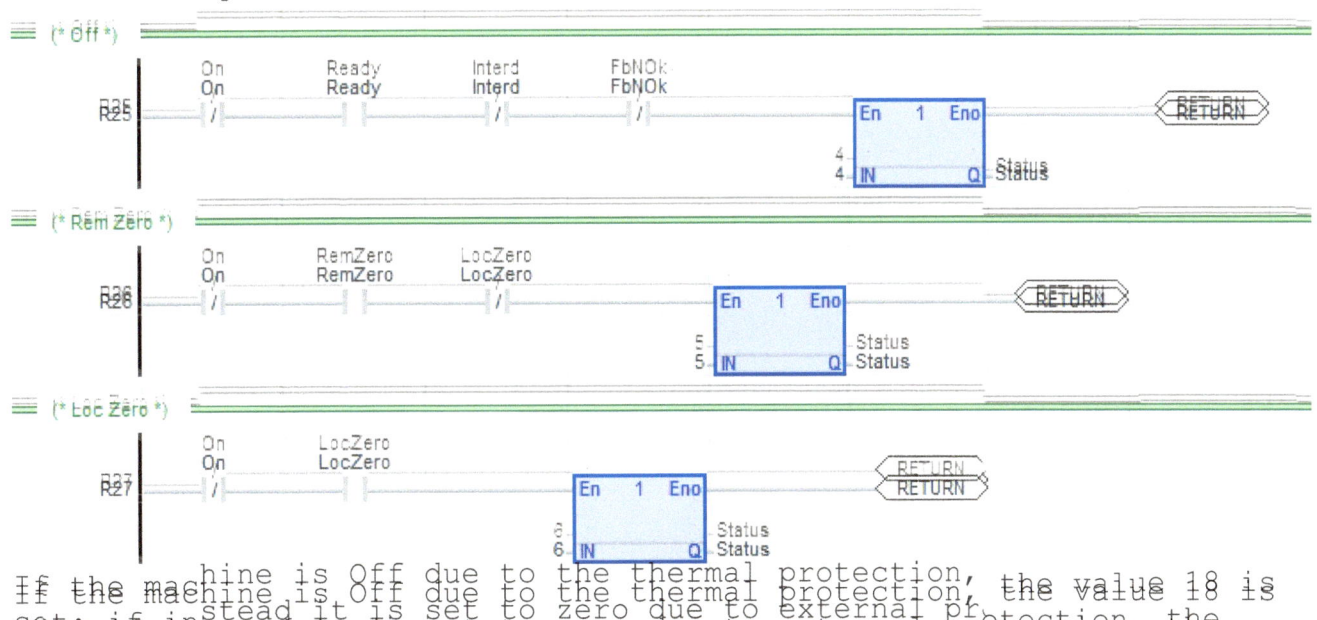

If the machine is Off due to the thermal protection, the value 18 is set; if instead it is set to zero due to external protection, the value 32 is set, as shown in figure 2.2.18:

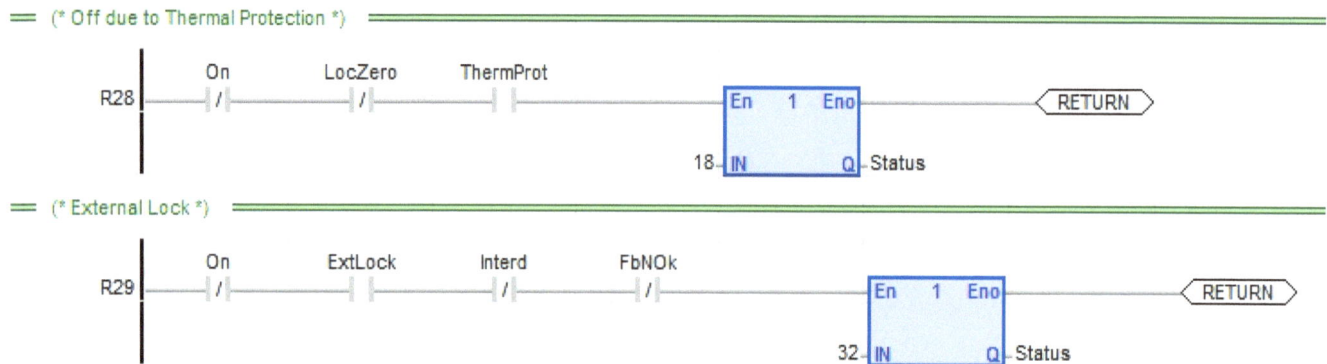

Finally, if the machine is Off due to the interdiction timer, the value 64 is set; if on the contrary it is Off in the presence of the non-status signal due to the external protection, the value 130 is set, as shown in figure 2.2.19. The justification for these seemingly strange values is given later when we will discuss the attributes to handle color in HMI text fields.

Note the presence of the RETURN statement at the end of each line. This instruction returns the logical flow to the calling program bypassing the execution of the rungs of subsequent instructions.

Map of instance variables

In summary, for each machine, six 16-bit registers of the PLC are required to be managed. We can use the first to aggregate all the Boolean input and output variables: ExtLock, On, ThermProt, rhReset and FbNok, Ready and Start. Two other registers are needed to store the remote and local RemCmd and LocCmd commands. The fourth register is required to store the Status output variable for decoding the operating status. The fifth and sixth register are used to store the number of starts and operating hours.

These registers will be declared in the tables of global and retentive variables; the RemCmd, LocCmd commands, start-ups and

operating hours must be declared in the retentive zone, so that they retain their values after a PLC restart.
Figure 2.2.20 below shows an example in the case of a booster system equipped with 4 electric pumps.

Retain Variables		
AI_Type	INT	%R254
areaTank	REAL	%R2020
CtTmrAlm	INT	%R318
Ep1BstCmdLoc	INT	%R13
Ep1BstCmdRem	INT	%R12
Ep1BstHH	INT	%R16
Ep1BstNr	INT	%R15
Ep2BstCmdLoc	INT	%R23
Ep2BstCmdRem	INT	%R22
Ep2BstHH	INT	%R26
Ep2BstNr	INT	%R25
Ep3BstCmdLoc	INT	%R33
Ep3BstCmdRem	INT	%R32
Ep3BstHH	INT	%R36
Ep3BstNr	INT	%R35
Ep4BstCmdLoc	INT	%R43
Ep4BstCmdRem	INT	%R42
Ep4BstHH	INT	%R46
Ep4BstNr	INT	%R45

2.3 The ElectricMotor Human Machine Interface

The local screen, on the HMI device, provides for the display, for each machine, of all the quantities acquired, described in previous chapter, as well as the possibility to locally control the start or stop of the machine with priority over the automatic control strategies. The operator will use this option in the event that one of the machines is malfunctioning and therefore goes out or in the commissioning phase of the control system if he wants to check the correct functioning of the automation chain, including electric wiring.

Remote supervision, on Scada systems, makes it possible to extend the functions of the HMI device, even to remote operators, connected via the Internet, which can completely acquire and command the machines themselves. A concrete example can be the remote control of a defrost sequence in the case of refrigeration systems.

The HMI device will show the values of the registers or in numerical form as in the case of the number of starts or operating hours or in the form of an alphanumeric string for remote and local commands and for the status.

The correspondence between numerical value and alphanumeric text string is decoded by specific tables configured on the HMI device.

Let's take as an example the typical display related to an electric pump inside a graphic page shown in figure 2.3.1. Starting from top to bottom we have three graphic objects: the status register display, the three-position local command button and the remote control register viewer.

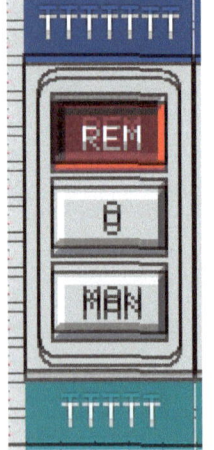

The status register graphic control is configured with the "Text Table Data Properties" dialog window. First of all, we must indicate in the Controller Register group the name of the variable that contains the status register: EpSts1 in our example. Subsequently, the Data Format must be defined, i.e. the justification of the text, the graphic Font, the number of alphanumeric digits, and finally the Text Table Number of decoding table, as shown in figure 2.3.2:

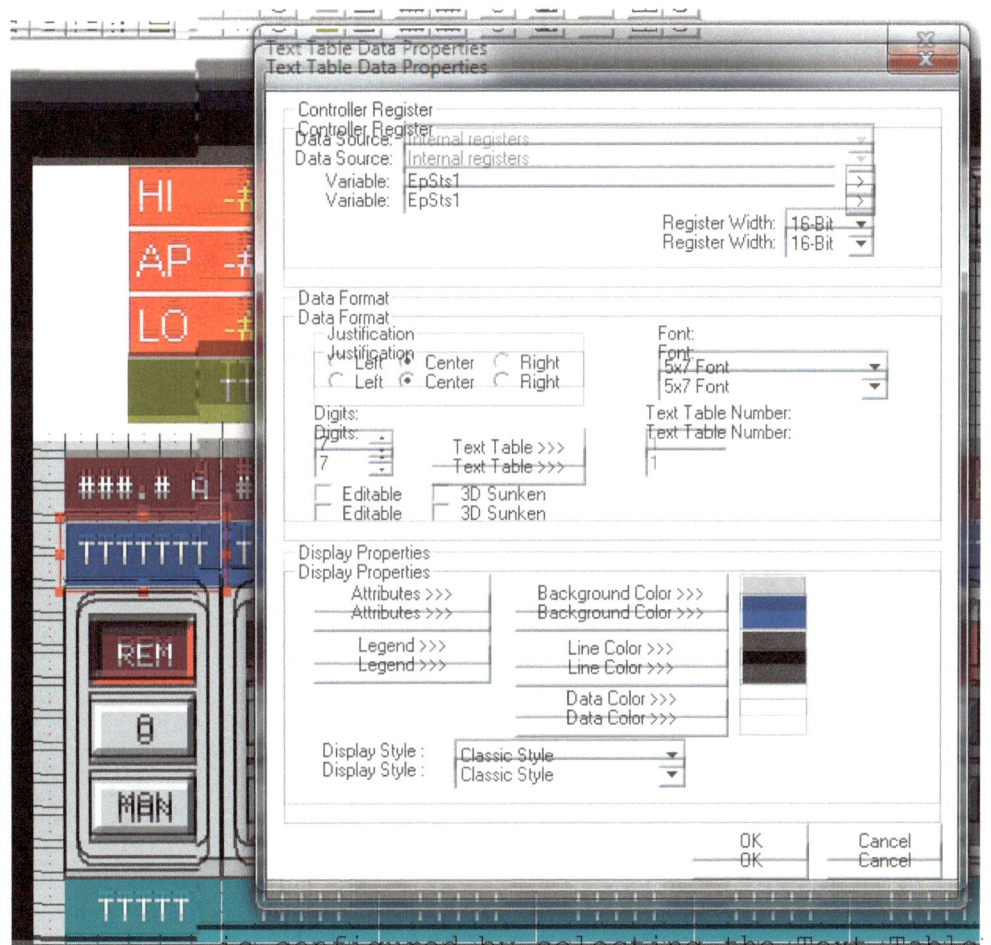

This table is configured by selecting the Text Table>>> button; on the new configuration window we enter, once a time with the Add button, the numerical values that the register can assume and the relative text that corresponds to this value. In our automation recipe we will specify, as shown in figure 2.3.3, the following value = text pairs, congruently with what has been defined in the control logic of the functional block illustrated in the previous chapter:

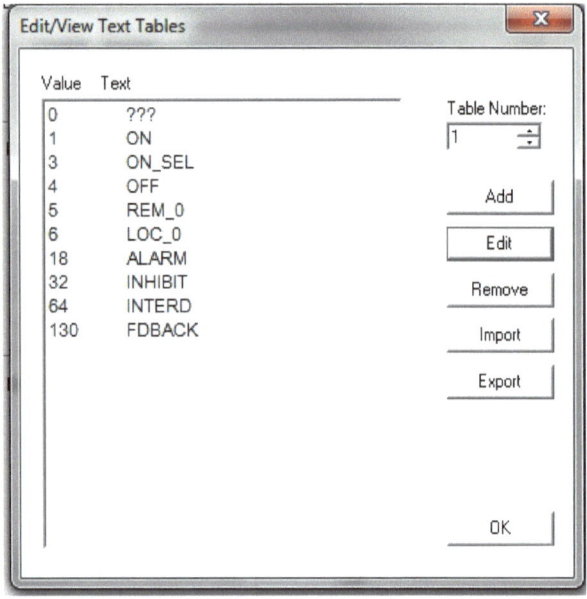

The meaning of the seemingly strange values such as 32, 64 etc. becomes clear when analyzing the settings of the Display Attributes group whose Override Register is set equal to the status register as shown in figure 2.3.4.

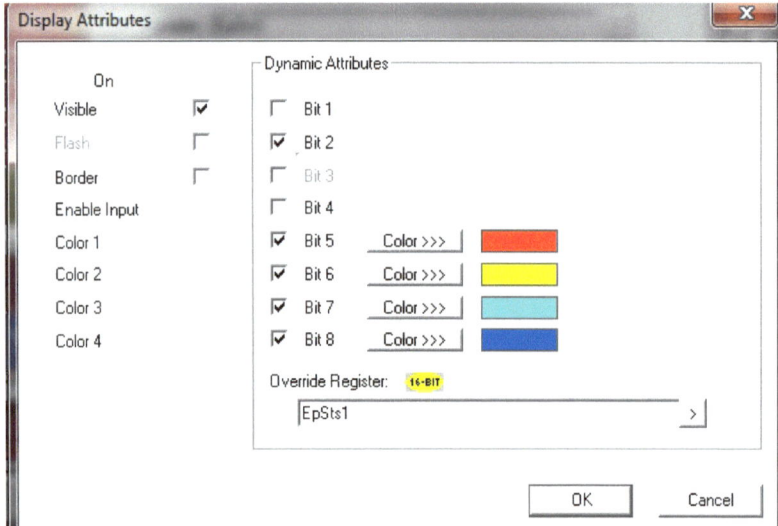

The value 3, corresponding to the ON_SEL status, On from the front-panel selector, produces the simultaneous activation of the bits 1 and 2 (3 in binary is written 00000011). The attribute Visible has no dynamic attributes because it is always On but the Flash attribute depends on Bit 2 causing the ON_SEL text to be displayed in flashing mode to attract the operator's attention.
The value 18, corresponding to the ALARM state, produces the simultaneous activation of the bits 2 and 5 (18 in binary is written

00010010) which corresponds to the dynamic attributes of Flash with background color Red.
The value 32, corresponding to the INHIBIT state, produces the activation of the Bit 6 with Yellow background color.
The value 64, corresponding to the INTERD state, produces the activation of Bit 7 with Cyan background color.
Finally, the value 130, corresponding to the FDBACK state, produces the simultaneous activation of the Bits 2 and 8, which corresponds to the dynamic attributes of Flash and Blue background color.
Similarly, table 2 decodes the RemCmd remote command register: the value 0 displays the text AUT; 1 the text STOP; 2 the text START as shown in figure 2.3.5.

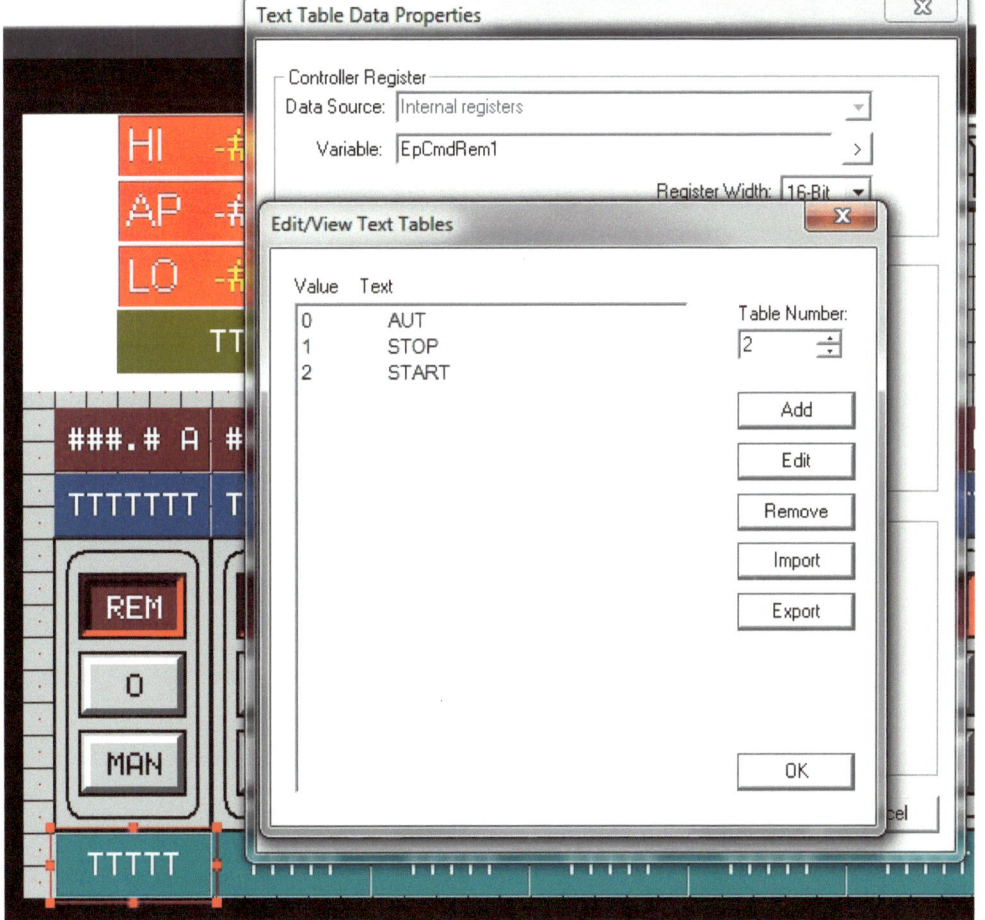

The three-position selector graphic object is used to enable the operator to send the mutually exclusive REM, O and MAN local commands corresponding to the values 0, 1, and 2 of the register named EpCmdLoc1 as shown in figure 2.3.6.

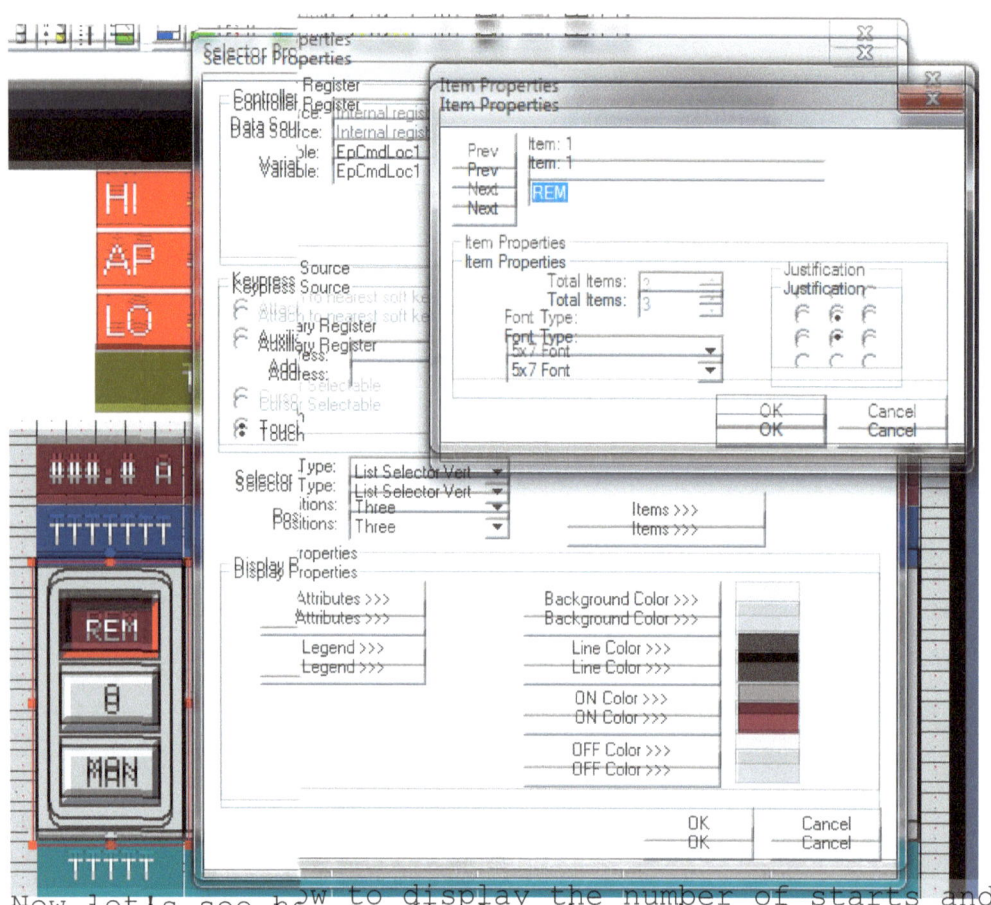

Now let's see how to display the number of starts and operating hours. The graphic object Numeric Data allows us to associate the Ep1CondHH variable by displaying it in a five-digit decimal format (max value is 30,000) and with engineering unit "h" as shown in figure 2.3.7:

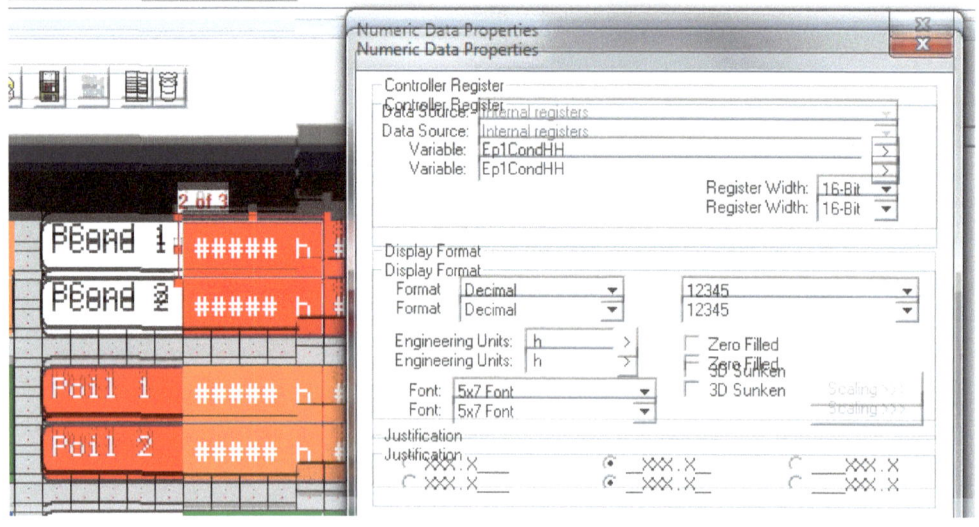

In a similar way the number of interventions is managed, but omitting to declare any engineering unit, as shown in figure 2:3:8:

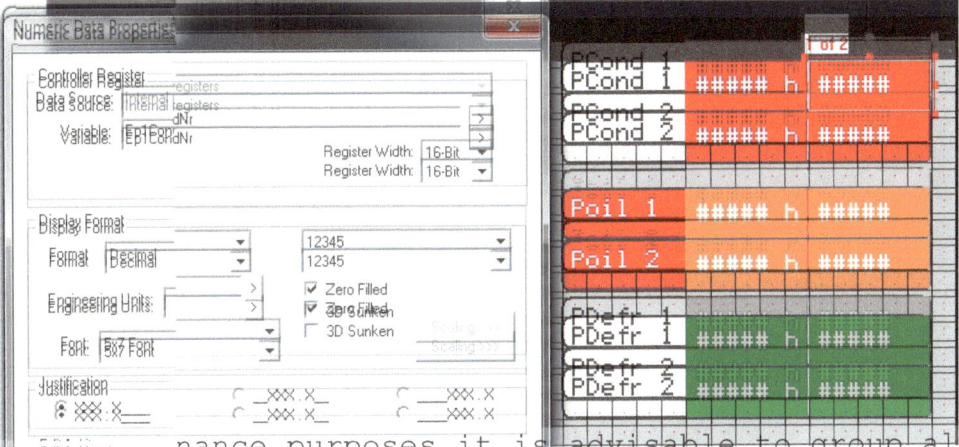

For maintenance purposes it is advisable to group all operating hours and the number of starts in a single HMI screen called HOURS. Figure 2:3:9 shows a real example taken from a refrigeration plant automation system.

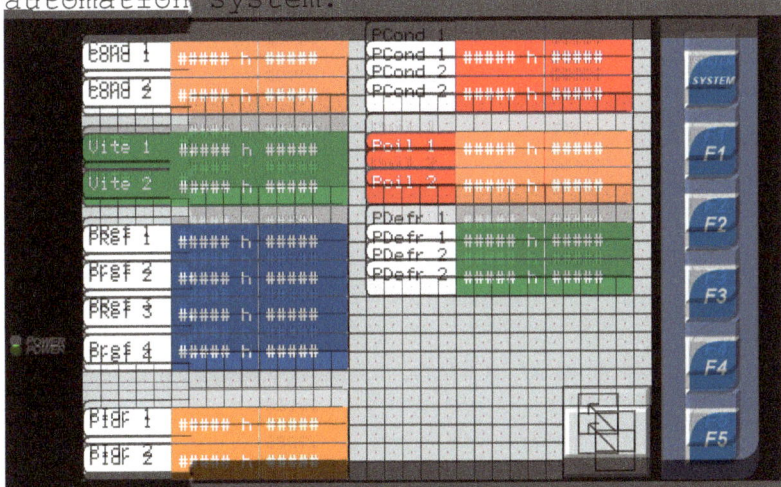

The description of the ElectricMotor pattern is now complete. In the next chapter we will present a functional block that allows two electric motors to be started in alternating sequence: the twin sequencer.

2.4 The twin sequencer function block

Problem

In water or HVAC systems, the presence of two twin pumps is very widespread; one of which has only the reserve function for the other. A classic example is a waste and / or meteoric water station with two pumps or a chilled / hot water secondary distribution line of a hydronic air conditioning system.

Solution

In order to make the wear equal, the pumps are generally started, one at a time, alternately, while in exceptional cases, corresponding to the exceeding of alarm thresholds, they can both be started or stopped. The Mot2Seq twin sequencer strategy dialogues with both ElectricMotor modules of the two pumps, sending them the start command via the Boolean Go variable and still receiving the information of On, Ready and FbNok. The control logic of the design pattern is contained in the UDFB module named Mot2Seq. The functional block will be called, within the main program, for each group of twin or twin pumps to be driven.

Map of local variables

The table of input, output and internal variables of the block is shown in figure 2.4.1:

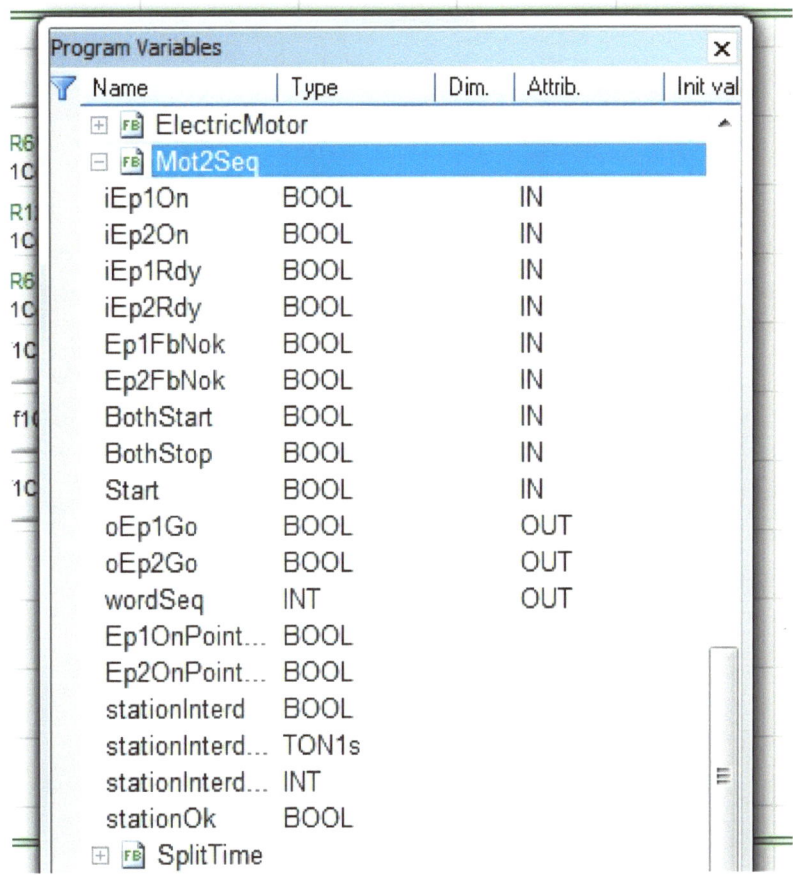

For each of the two controlled pumps the Boolean variables of On, ready to start (Rdy), and no feedback state (FbNok) are acquired. There are also three other Boolean input variables that respectively indicate: BothStart the need to start both pumps, BothStop to stop both, and finally Start indicating the need to activate the twin sequence strategy or not.

The output variables are three: the two Boolean variables of Go for the enabling to the start for each pump and the whole 16-bit wordSeq variable that indicates the status of the sequencing logic, used to monitor the correct functioning of the sequencer.

An example of EpCondSeq instance, recalled by the main program, for the alternation of two condensation pumps in a refrigeration plant, is shown in Figure 2.4.2:

The internal logic of the UDFB module is shown in the following rungs:

Rungs R1=R2, as shown in figure 2.4.3, reset the stationInterd interdiction flag by means of a standard delayed timer. This timer will be activated, via the flag, every time the strategy takes any action on the controlled pumps. Setting this timer to 10 seconds ensure us that these actions can not be taken too frequently. If desired, the block can be easily modified by adding the value of the interdiction timer set to the input parameters.

The rungs R3=R4 command, in the presence of the external signal of BothStart, the Go of both pumps, also setting to 2 the value of the wordSeq and returning the logic flow to the main program, as shown in figure 2.4.4:

The R5-R6 rungs control, in the presence of the external signal of BothStop or that of lack of Start, the stop of both pumps (resetting the relative Go), also setting to 3 the value of the wordSeq and returning the logic flow to the main program, as shown in figure 2.4.5.

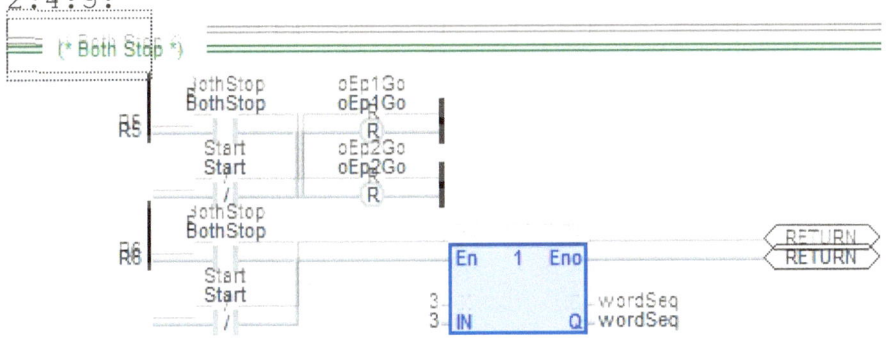

The R7 rung, in the presence of an interdiction flag, sets the wordSeq to 1, returning the logic flow to the main program, as shown in figure 2.4.6:

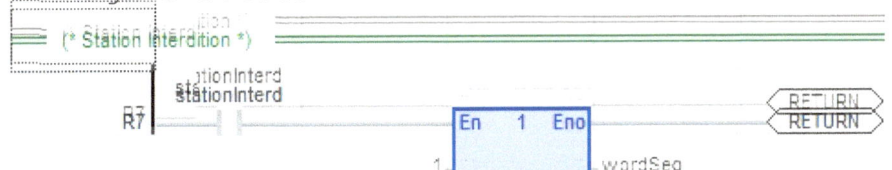

The R8-R9 rungs verify that a pump is started and not in a failed feedback state while the other is stopped. In this case, the block recognizes that everything is OK and therefore sets the value of the wordSeq to 0 and returns control to the main program, as shown in figure 2.4.7:

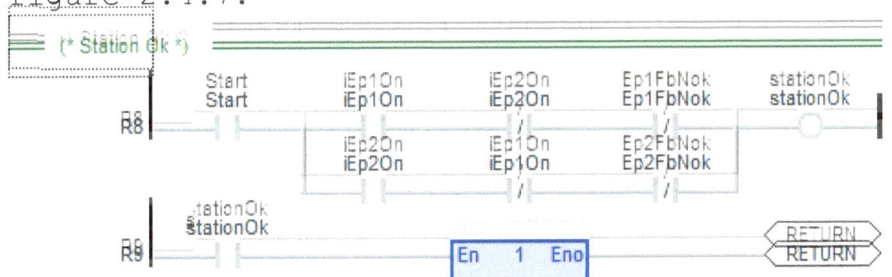

If rung R9 is not satisfied, the logic flow continues with rung R10 which, if pump start pointers are found both set to 0, set immediately the first starting point, as shown in figure 2.4.8:

Rung R11 checks whether there are the conditions to start the first pump, i.e. if first pump is On pointed and it is ready to start and not in a FeedBackNotOk status. In this case, the Go command is given to the first pump, while the consent to the second is denied, and finally the On pointers are updated and the interdiction flag set. Rung R12 sets the value of wordSeq to 4 if the previous line has energized the stationInterd flag and returns the flow to the calling program; otherwise the rung R13 update the pointers of the pumps to be started next, as shown in figure 2.4.9.

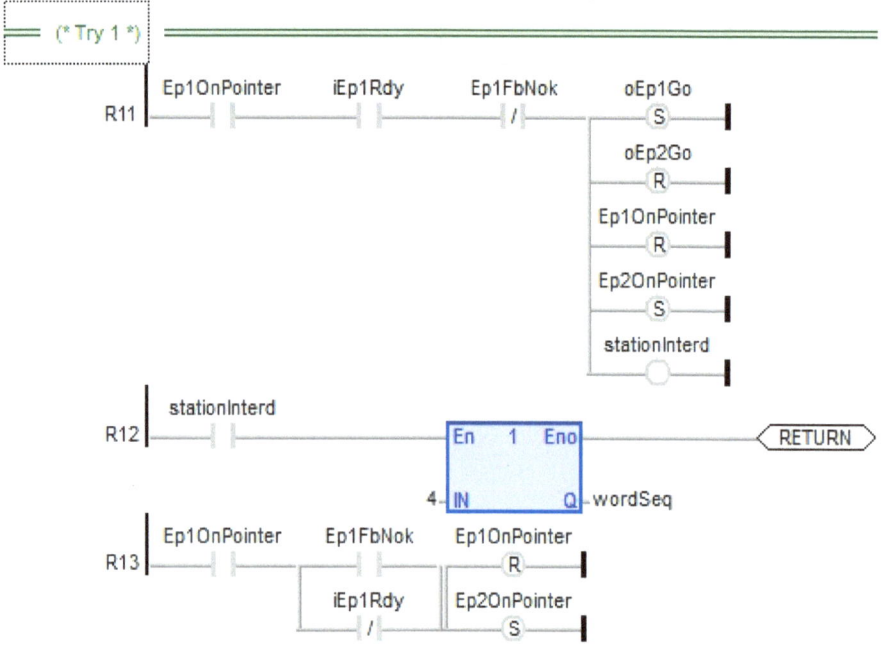

The R14-R16 rungs operate like the R10 and R12 rungs but on the second pump instead of the first one, as shown in figure 2.4.10.

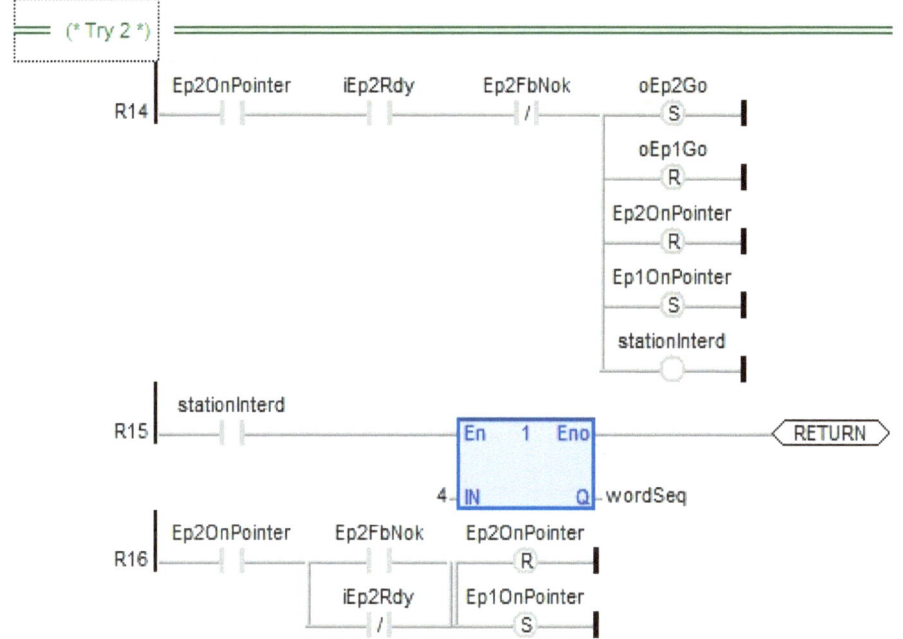

The twin-pump sequencer function block is now fully implemented.

2.5 The twin sequencer Human Machine Interface

An example of a screen displaying a pumping station with a twin pump sequencer is shown in figure 2.5.1:

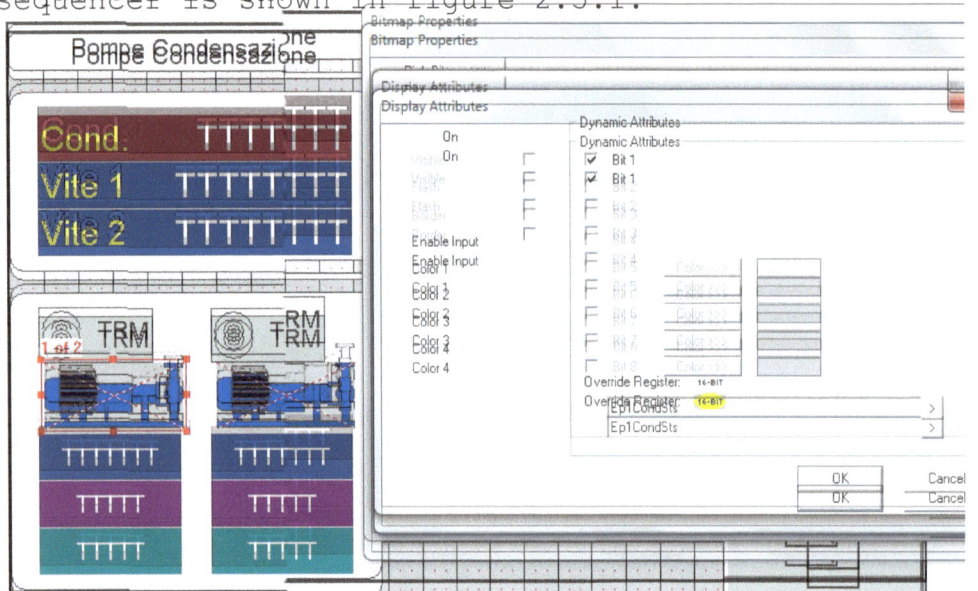

The pump symbol becomes visible only if the pump is On; only in this condition the bit 1 of the status variable associated with the pump is active (see block UDFB ElectricMotor):

3. Summary

We have reached the end of our first booklet. I would like to thank the reader for the effort made to read it and for the trust given to me as the author.

I am sure that the result achieved will be fully positive and that the quality of the future work carried out, by implementing the techniques acquired, will emerge with sufficient clarity and will be a source of great professional satisfaction.

A sincere wish of good work and a goodbye to the next booklet that will deal with the automation of 4-20 mA transmitters:

www.ingramcontent.com/pod-product-compliance
Lightning Source LLC
Chambersburg PA
CBHW051935210526
45473CB00006B/2251